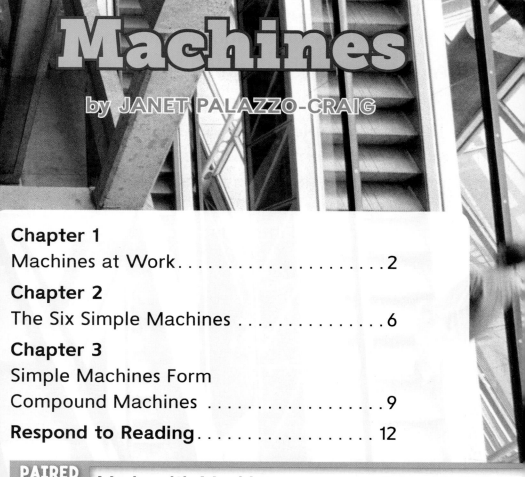

Genre Nonfiction

W9-AWH-971

Essential Question

How do machines make our lives easier?

Machines

by JANET PALAZZO-CRAIG

Chapter 1
Machines at Work

The world is a busy place! Look around you. People are riding on skateboards, driving cars, climbing stairs, lifting packages, and digging in gardens. In each case, some kind of work is being done.

People have developed special tools that help you do work. These tools are called machines.

There are **simple machines** and **compound machines**. Simple machines are the most basic kinds of machines. These machines have few or no moving parts. Compound machines are two or more simple machines combined into one machine.

A skateboard is a compound machine made up of simple machines.

Each person is doing work.

Work It!

Doing work means moving something or changing its motion. When you clean your room, you move dirty clothes from the floor to a laundry basket. When you take out the garbage, you move bags to a trash can.

For work to be done, some effort, or force, has to be used. Your muscles provide the effort for simple machines when you push or pull. The effort for compound machines may also come from your muscles. It can come from other sources, as well, including wind or electricity.

A ramp can help you do work.

How Do Machines Help?

Machines help you do work. A simple machine does not really reduce the work it takes to do a job. It simply lets you use less force. However, there is a trade-off. When you use less force, you have to apply that force over a greater distance.

Imagine lifting your bike up and into a truck. Lifting takes hard work. Now imagine placing a ramp from the ground to the truck. Instead of lifting the bike up, you can just push or pull it up the ramp.

The ramp is a simple machine called an **inclined plane**. The push or pull you apply to move the bike is called the **effort force**. You use less effort force when you use the ramp. Although the effort force is less, you must move the bike a longer distance than you would by lifting the bike up. This means there is a trade-off between the amount of force used and the distance moved to do the work.

In each case, you actually do the same amount of work. By using the ramp instead of lifting, the work seems easier because you use less effort.

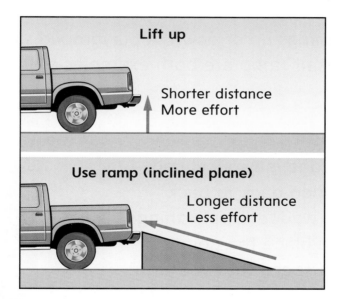

Lift up

Shorter distance
More effort

Use ramp (inclined plane)

Longer distance
Less effort

Using a ramp (inclined plane) to move something takes less effort force than lifting it even though the distance you move it is longer.

Chapter 2
The Six Simple Machines

There are six kinds of simple machines. The six machines are the inclined plane, the **lever**, the **wheel and axle**, the **pulley**, the **screw**, and the **wedge**. All machines are made of some combination of these simple machines.

The Inclined Plane

The inclined plane is the simplest of the simple machines. It has no moving parts and is a flat, slanting surface. An inclined plane makes it easier to move things up or down. A playground slide and a wheelchair ramp are both examples of inclined planes.

The Wedge

The wedge is two inclined planes joined back to back. You use a wedge to split things apart. You place the pointed end into a space that you want to open and push on the wide end. The wedge changes the downward force into a sideways force. The blade of an ax is a wedge.

Wedge

The Screw

The screw is really an inclined plane. It has a slanted surface wrapped around a center pole. This slanted surface is an inclined plane. It lets the screw move forward when it is turned.

The Lever Gives You a Lift!

The lever helps you lift or open things. This simple machine is a bar that rests on a fixed point, called a fulcrum. The fulcrum lets the bar pivot, or turn. When you use a lever, you apply an effort force. The object you are moving is called a load. A bottle opener is a lever, and so is a shovel.

Screw

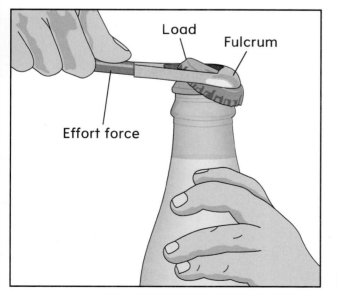

Load

Fulcrum

Effort force

Lever

Pulling with the Pulley

The pulley is a wheel with a rope or chain around it. It helps you lift things by changing the direction of your force. The pulley works because pulling down seems easier than lifting up. To use a pulley, you attach a heavy load to one end of the rope. You pull down on the other end, and the heavy load lifts up.

A pulley makes it easier to lift heavy objects.

Rolling with the Wheel and Axle

You can think of the wheel and axle as a lever that rotates in a circle. A wheel is connected to a rod called an axle, and the wheel turns with the axle. This simple machine makes it easier to move or turn objects.

The wheel is larger than the axle. As they turn together, the wheel takes less effort force to move than the axle does, but it moves a longer distance. The axle takes more force to move, but it moves a shorter distance.

A doorknob is a wheel that connects to a rod, which is an axle. You could open the door by turning the knob or the rod. The knob takes less effort force to turn than the rod does, which makes opening the door easier. You must turn the knob a longer distance than the rod, though.

Chapter 3
Simple Machines Form Compound Machines

When two or more simple machines work together, they form a compound machine. Bicycles, hand drills, and escalators are some examples of compound machines you may use.

The Can Opener Can Do It!

Did you know that a can opener is a compound machine? It is made up of these simple machines working together:

- Lever (The hinged handle)

- Wheel and axle (The turning knob)

- Wedge (The sharp blade that cuts the metal can)

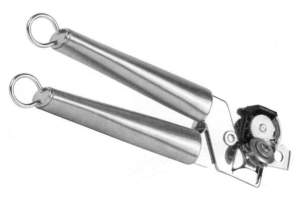

A can opener is a compound machine made up of simple machines that work together.

The Escalator: Step Right Up!

Have you ever been to a shopping mall and taken a ride up or down the escalator? If so, you have used another compound machine. An escalator takes the work out of walking up or down stairs. An escalator is powered by an electric motor.

These simple machines make up an escalator:

- Inclined plane
- Wheels and axles
- Pulleys

How do these simple machines work together to make the escalator run? First, think about the slant of the steps, which form an inclined plane.

Simple machines work together in this compound machine.

Next, take a look inside the escalator to see where the wheels and axles and pulleys are located. You'll see something that looks like the chain on a bicycle. A group of wheels is stretched around two large gears below the escalator's steps. Gears are wheels with teeth on them. The teeth are small wedges that can move other gears or objects as they turn.

As the escalator's gears turn, they move the wheels in a circle under the steps. This motion moves the steps up or down an inclined plane. The electric motor is attached to one of the large gears with a pulley. As a pulley turns the large gear, the wheels under the steps are set in motion. The wheels and gears work like a wheel and axle. Pulleys also move the handrail of the escalator.

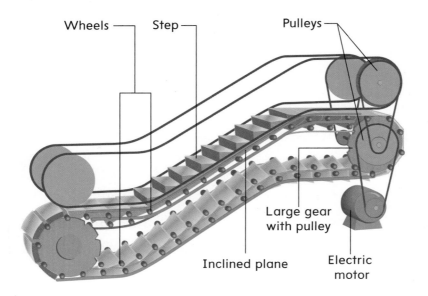

Wheels — Step — Pulleys —

Large gear with pulley

Inclined plane

Electric motor

Respond to Reading

Summarize

Use details found in the text to summarize *Machines*. The graphic organizer may be of help to you.

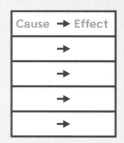

Text Evidence

1. What are simple machines? What are compound machines?

2. Working with a partner, read the book again. Discuss how using simple machines helps you do work. This is the effect. As a team, fill in a cause-and-effect chart to show the causes. CAUSE AND EFFECT

3. Use what you know of antonyms to determine the antonym of *same* on page 5. ANTONYMS

4. Write a letter to a movie theater owner. Try to persuade the owner to put in a wheelchair ramp. Explain the effects of putting in the ramp and why you think it is important. Include words to persuade, such as *should* and *must*. WRITE ABOUT READING

Compare Texts

Read about how Pedro and Mia used simple machines to help build a playground.

Made with Machines

Pedro and Mia were excited; they were going to help to build a neighborhood playground.

"Can you kids bring the boards up here?" called one man standing high up on a platform.

"We can do it!" Pedro responded.

Pedro and Mia tried carrying the boards, but Pedro sighed. "There's no way we can carry these across the playground."

"Let's use the wheelbarrow," Mia suggested.

Using the wheelbarrow, it was easy to move the boards. How, they wondered, could they possibly heave the heavy boards up to the platform?

"Put the boards in the big bucket, and I'll haul on the rope," the man told them.

The children put the boards in the bucket, the man pulled on the rope, the bucket rose into the air, and soon the boards were on the platform.

"We'll bring the paint cans," Pedro called out eagerly. The cans were heavy, though, and the wheelbarrow could not be rolled up stairs.

Pedro placed a plank on top of the stairs. Pedro and Mia rolled the wheelbarrow up the ramp. Soon, they were painting the swing set.

Finally, the playground was finished. "We couldn't have done it without our machines," beamed Pedro.

Make Connections

Explain how Pedro and Mia used machines as they worked on the playground. TEXT TO TEXT

Glossary

compound machine *(KOM-pownd muh-SHEEN)* a combination of two or more simple machines *(page 2)*

effort force *(EF-uhrt FORS)* a push or pull you apply to a simple machine to move something *(page 5)*

inclined plane *(in-KLINED PLAYN)* a simple machine formed by a flat, slanting surface *(page 5)*

lever *(LEV-uhr)* a simple machine made of a bar and a fixed point, called a fulcrum *(page 6)*

pulley *(PUL-ee)* a simple machine made up of a rope or chain wrapped around a wheel *(page 6)*

screw *(SKREW)* a simple machine that is an inclined plane wrapped into a spiral *(page 6)*

simple machine *(SIM-puhl muh-SHEEN)* a machine with few or no moving parts, making it easier to do work *(page 2)*

wedge *(WEJ)* a simple machine that uses force to split objects apart *(page 6)*

wheel and axle *(WEEL and AK-suhl)* a simple machine made of a rod called an axle attached to the center of a wheel *(page 6)*

Index

Focus on Science

Purpose To show how a ramp is a simple machine

Procedure

Step 1 Will the steepness of a ramp affect how much force is needed to lift an object? Make a prediction.

....

Step 2 Create a ramp using several books and a board. Use a spring scale to measure the amount of force needed to lift a book tied with string straight up the ramp. Record your results in your science journal.

....

Step 3 Use the spring scale to slowly pull the book up the ramp. Record your results.

....

Step 4 Share your comic strip with your classmates.

Conclusion Did the amount of force needed to pull the book up the ramp change when you made the ramp steeper? Explain your answer. Perform this experiment again, using objects of different masses.

Machines | Science

Lexile 790

www.mheonline.com/inspire-science

978-0-02-133476-6
MHID 0-02-133476-5

EAN

9 780021 334766

99701

3

Mc
Graw
Hill
Education

Magnets Attract!

by Meish Goldish

PAIRED READ Lost!

STRATEGIES & SKILLS

Comprehension
Strategy: Summarize
Skill: Compare and Contrast

Vocabulary Strategy
Context Clues

Vocabulary
atom, attract, compass, inner core, magnetic, magnetic field, outer core, pole, repel

Content Standards
Science
Physical Science

Word Count: 1049**

Photography Credit: Cover Richard Newstead/Getty Images

**The total word count is based on words in complete sentences found in the running text, sidebars, headings, and captions. Numerals and words in phrases that comprise labels, diagrams, and headings are not included.

ConnectED.mcgraw-hill.com

Send all inquiries to:
McGraw-Hill Education
8787 Orion Place
Columbus, OH 43240

ISBN: 978-0-02-135922-6
MHID: 0-02-135922-9

Printed in the United States of America.

2 3 4 5 6 7 8 9 LCR 20 19 18 17

A

Genre Nonfiction

Essential Question
How do magnets work?

Magnets Attract!

by Meish Goldish

Chapter 1
What Is a Magnet?

Most magnets are made of iron. A thing a magnet can pull is **magnetic**. Magnetic things are made of iron or other metals. A magnet does not **attract** things like cloth or wood. They are not magnetic.

Magnets have a strong pull. How does it work? All of the **atoms** in a magnet line up and face the same way. Some things are made of jumbled atoms that do not line up. They are not magnets.

Magnets hold things to metal surfaces.

What happens if you put two magnets side by side? Do you think they pull on each other? Some times they will. Other times they will not.

Magnets have two ends. Each end is a **pole**. One end is the north pole. The other end is the south pole. How the poles of the magnets are placed tells what will happen.

Poles that are not the same pull on each other. What do you get if the north pole of one magnet faces the south pole of another magnet? The two magnets pull on each other.

What happens if two north poles or two south poles are facing each other? Then the two magnets **repel**, or push on each other. What is the rule? Poles that are not the same attract, but poles that are the same repel.

Chapter 2
The World's Largest Magnet

Some magnets are square. Magnets can look like round flat wheels. Some look like horseshoes. A magnet can be any shape or size.

What is the world's largest magnet? Do you know what it is? It is Earth!

Earth has an **inner core** and an **outer core**. They are made of iron. The iron turns in the cores. This makes Earth's magnetic field. A **magnetic field** is the space near a magnet where its pull can be felt. All magnets have a magnetic field.

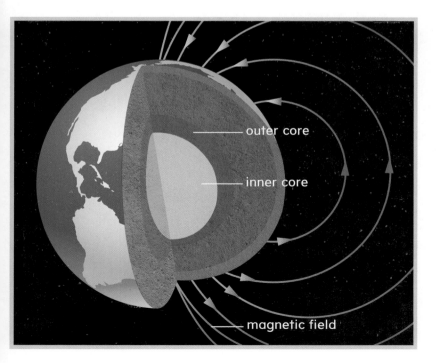

outer core

inner core

magnetic field

Earth's magnetic field is what makes a **compass** work. A compass has a thin magnetic strip of iron. Earth's magnetic field points the north pole of the strip north. It points the south pole of the strip south. A compass can tell you which way you are looking.

A compass can help you to understand a map.

Chapter 3
Magnets for Travelers

Long ago, people on a trip used magnets to find their way. They used a stone called magnetite. It is a natural magnet. It has a lot of iron in it.

On ships, magnetite was hung from a string. The stone pointed north and south. That told which way the ship sailed.

Magnetite was an early compass. It was called a lodestone.

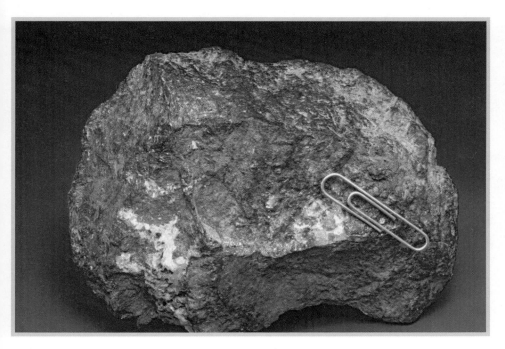

A piece of magnetite—a natural magnet

Some animals have a natural compass in their bodies. Birds that migrate, like geese, can feel Earth's magnetic field. It tells them which way to go. Whales, sea turtles, and salmon also use Earth's magnetic field. It guides them through vast spaces of water.

Chapter 4

How You Use Magnets

You use magnets all the time. Some can openers have small magnets to catch the lid of the can. A crane has a large magnet to lift piles of metal junk.

Magnets are used in things that have electric motors. There are magnets in CD players, blow dryers, and cars.

Magnets are also found in most things that have speakers.

This electric can opener works with a small magnet.

This huge electromagnet picks up scrap metal.

Turn on a computer. A hard drive starts to spin. Its hard disk has a magnetic coating that holds data. The drive's head can read the magnetic fields on the disk. The data are sent to other parts of the computer.

Look at the black strip on the back of a credit card. It is a magnetic stripe. It is a magstripe. It is made of magnetic bits. To pay for something, people swipe their cards through magstripe readers.

Computer hard drive

The black bar on the back of a credit card is magnetic.

Doctors use magnets to see what is in a patient's body. They use a machine called an MRI (magnetic resonance imager). It has a powerful magnet. The patient lies in the machine. The MRI makes a strong magnetic field. It makes atoms in the body act like magnets. The machine can then record body images. They look like X-ray pictures.

Images taken with an MRI

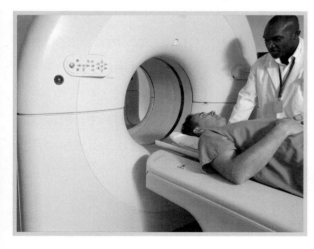

An MRI uses magnets to see inside a patient's body.

Respond to Reading

Summarize

Write a summary of what you read. Give details from the text. The graphic organizer may help you.

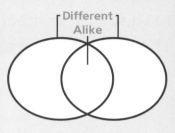

Text Evidence

1. Why aren't all things magnets?

2. Read the book with your group. Compare magnetic and nonmagnetic materials. COMPARE AND CONTRAST

3. What context clues helped you understand the word *jumbled* on page 2? CONTEXT CLUES

4. Suppose you explore with magnets. What happens when you put a magnet next to a paper clip? What happens when you put a magnet next to a piece of paper? Tell if each is attracted, repelled, or not moved by the magnet. Compare your work with a partner's. WRITE ABOUT READING

Compare Texts

Read how science helps a boy determine the right direction to travel in the fog.

Lost!

"What are we going to do?" Nick thought. Nick and his father were out in a boat, fishing on the bay. Fog rolled in as thick as smoke. It was impossible to see the shore. The motor had stalled. Nick's dad was looking at it. Even if he fixed the motor, they needed to know which way to go. His father hadn't brought a compass. A compass tells which way to go.

Nick emptied his pockets onto the boat deck. He found a magnet, a rock, a penny, and a piece of cork. He pushed aside the penny and the rock.

Nick picked up the magnet and the cork. He had learned how to make a compass with a magnet. Could he remember how? And if he did, would it work? Excitedly, Nick found a needle in the first aid kit in the cabin. He stroked the needle across the north end of the magnet in one direction. He counted 50 strokes. Then he pushed the needle into the cork. Nick floated the cork in a bowl of water. He held his breath. The needle moved. It aligned with Earth's magnetic field, pointing north to south.

Nick let out a sigh of relief. Just then, the motor started. Nick's dad saw the compass. "I'm proud of you, son," he said. Once they knew which way to go, the boat took off.

 Make Connections
What did Nick need to know about how magnets worked? TEXT TO TEXT

Glossary

atom *(AT-uhm)* the smallest particle of an element *(page 2)*

attract *(uh-TRAKT)* to pull something *(page 2)*

compass *(KUM-puhs)* a magnetic tool for showing direction *(page 6)*

inner core *(IN-uhr KAWR)* a sphere of solid material at the center of Earth *(page 5)*

magnetic *(mag-NET-ik)* able to be pulled to a magnet *(page 2)*

magnetic field *(mag-NET-ik FEELD)* the space around a magnet where its pull can be felt *(page 5)*

outer core *(OW-tuhr KAWR)* a liquid layer of Earth lying below the mantle *(page 5)*

pole *(POHL)* one of two ends of a magnet where the pull of the magnet is stronger *(page 3)*

repel *(ri-PEL)* to push or force away *(page 4)*

Index